YOUR KNOWLEDGE HAS VALUE

- We will publish your bachelor's and master's thesis, essays and papers

- Your own eBook and book - sold worldwide in all relevant shops

- Earn money with each sale

Upload your text at www.GRIN.com and publish for free

Ieva Bekeryte

Satellite systems - History, definition, functioning principles and application spheres

GRIN Verlag

Bibliografische Information der Deutschen Nationalbibliothek:

Die Deutsche Bibliothek verzeichnet diese Publikation in der Deutschen National-
bibliografie; detaillierte bibliografische Daten sind im Internet über http://dnb.d-
nb.de/ abrufbar.

Dieses Werk sowie alle darin enthaltenen einzelnen Beiträge und Abbildungen
sind urheberrechtlich geschützt. Jede Verwertung, die nicht ausdrücklich vom
Urheberrechtsschutz zugelassen ist, bedarf der vorherigen Zustimmung des Verla-
ges. Das gilt insbesondere für Vervielfältigungen, Bearbeitungen, Übersetzungen,
Mikroverfilmungen, Auswertungen durch Datenbanken und für die Einspeicherung
und Verarbeitung in elektronische Systeme. Alle Rechte, auch die des auszugsweisen
Nachdrucks, der fotomechanischen Wiedergabe (einschließlich Mikrokopie) sowie
der Auswertung durch Datenbanken oder ähnliche Einrichtungen, vorbehalten.

Imprint:

Copyright © 2006 GRIN Verlag GmbH
Druck und Bindung: Books on Demand GmbH, Norderstedt Germany
ISBN: 978-3-638-73576-6

This book at GRIN:

http://www.grin.com/en/e-book/75050/satellite-systems-history-definition-functio-
ning-principles-and-application

GRIN - Your knowledge has value

Der GRIN Verlag publiziert seit 1998 wissenschaftliche Arbeiten von Studenten, Hochschullehrern und anderen Akademikern als eBook und gedrucktes Buch. Die Verlagswebsite www.grin.com ist die ideale Plattform zur Veröffentlichung von Hausarbeiten, Abschlussarbeiten, wissenschaftlichen Aufsätzen, Dissertationen und Fachbüchern.

Visit us on the internet:

http://www.grin.com/

http://www.facebook.com/grincom

http://www.twitter.com/grin_com

Hochschule Reutlingen
Reutlingen University

Satellite Systems

Advanced Communications - Master 2nd Semester

Fachhochschule Reutlingen

Written by: Ieva Bekeryte

Submission date: 28.06.2006

Table of contents

List of abbreviations

BER	Bit Error Rate
DBS	Direct Broadcast Satellite
DTH	Direct – to – Home
EHF	Extremely High Frequency
FSS	Fixed Service Satellite
FTA	Free – to – Air
GEO	Geostationary satellite
G_r	Receiver gain
G_t	Transmitted antenna gain
HEO	Highly Elliptical Earth Orbit
HPA	High Power Amplifier
IF	Intermediate Frequency
ITU	International Telecommunication Union
L_a	Antenna loss
LEO	Low
L_l	Line loss
LNA	Low – Noise Amplifier
L_s	Space loss
MEO	Medium Earth Orbit
MPEG	Moving Picture Experts Group
RF	Radio Frequency
Satcom	Satellite Communication
SHF	Super High Frequency
SNR	Signal to Noise Ratio
Ts	Noise temperature
TT&C	Tracking, Telemetry & Control

TWTA	Traveling Wave Tube Amplifier
UHF	Ultra High Frequency
VSAT	Very Small Aperture Terminal

List of figures

List of tables

1   Introduction

One of the most important success factors today is quick and perfect communication. Such a communication is quite easy to ensure, when the communicators are near each other, but the difficulties arise as the distance increase. Already many years ago people used various means of communication between longer distances (smoke signals, sounds, different colors flags and etc.).

The age of wire – based communication began after the A. G. Bell first time transmitted his voice in 1875. The following inventions presented the two new ways of communication and networking, which are fiber-optic and satellite communications. Both opened the gates to innovative types and dimensions of services. Nevertheless the commonly known and used satellites' function to broadcast television programs, many other services can be provided via these transmitters in the space. Nowadays satellites are used for various different reasons, such as weather forecasts, navigation systems, observation of other planets, scientific researches, military purposes, communication and etc.

Satellites and their systems play the significant role in our every day life and are used in places where it was even not imagined that they can be adapted (e. g. press).

Without satellites our life would be considerable different. Thus, the goal of this paper is to present this great invention of XX age, which enabled the society to speed – up the communication even more than before and opened the door to many new discoveries. The main focus is going to be concentrated on communication satellites and their main working principles as they are a part of nowadays used means of advanced communication.

1.1   Structure of Project

Firstly, the basic facts of satellites, including history, definition, types of satellites, their orbits and advantages as well as disadvantages, will be presented. Afterwards, the main functioning principles of communication satellites will be described in order to introduce the most important aspects, how the messages are transmitted. Finally, the application spheres of communication satellites will be listed and described more in detail.

2 Basics Facts

The background information about satellites is presented in this chapter, which is essential in order to understand their benefits for the communication of the world.

2.1 Definition and History

Usually, satellite is defined as a natural space body orbiting around the other natural space body. As for example, moon is the satellite of earth and earth is the satellite of sun. Nowadays the artificial spacecrafts are named satellites as well, because they are launched by a rocket to the space and are kept there by gravitational force revolving around the earth similarly as planets can orbit around the other planets. Satellites are highly specialized wireless receivers/transmitters, which main function is to relay the radio frequency waves and the encoded information in them from one corner of the world to another. Currently there are hundreds operating satellites above the earth.

The science fiction writer Arthur C. Clarke from England could be name the father of satellite communications, because he was the first who proposed to launch a satellite into the Earth orbit where satellite's speed would match with the rotation of the Earth. That orbit, which is 35786 km height above the planet surface today, is known as geostationary orbit, but sometimes it is called Clarke orbit in honor of his work and ideas [Held91]. 1945 Mr. Clarke was analyzing different orbits and was stressing the possible high-speed global communication networks enabled by above the earth surface revolving satellites. Already at that time, he emphasized that it would be enough to have 3 satellites in order to cover the whole planet. The first world's satellite Sputnik 1 as big as basketball was launched by Russia (former Soviet Union) in 1957 (October) with the aim to relay the signal of Morse code [Wiki06b].

The first commercial satellite PAS – 1 was launched by American satellite operator PanAmSat in 1984 as the governmental providers were able to offer quite poor services at the high cost.

2.2 Registration of satellites

Not all the countries are capable to launch satellites by themselves though lots of can design and build them, because of higher costs and required knowledge about launching. Recently various conjunctions (EU) and splits (Soviet Union) were made between different countries, which lowered the number of launching capable countries (Soviet Union, USA, EU, Israel and etc.) [Wiki06b].

Similar as all the ships should be registered in central register, satellites has to be recorded as well. The registration process starts in the own country (where launched or produced - I think produced), e. g. in Germany responsible is Bundesamt für Luftfahrt (Braunschweig) (approx. 100 satellites were registered till 1997) and the company forwarded the registration further to UNO in New York, which list the satellites of the entire world. Speaking only about satellites, which broadcast TV and radio signals, there are more than 100 such a communication satellites today placed in the space. Much more complicated taking much more time is the transmission frequency (Funkanmeldung) registration, which is made by in scope of United Nations working agency called International Telecommunications Union the Radiocommunication department (ITU - R), located in Genf, in Switzerland. The main responsibility of this organization is to ensure the possibility to transmit and administer the limited resources of orbits in space as well as the spectrum of radio frequency (RF). Thus, the orbit position, the frequency spectrum from 9 kHz till 1000 GHz (terrestrial and via satellite) and also the wire line connected communication is in charge of ITU organization. It regulates and administrates everything, what can radiate starting from garage door opener and ending with high capacity radar. Registrations are made through signatory states (eg. In Germany: RegTp = Regulierungsbehörde für Telekommunikation und Post) according principle "First come, first Served". The first step of the registration of transmission equipment (Funkanlage) is the so called Advanced Publication, where the operator should describe for the society the characteristics of his new service, such as desired orbit position, frequency band, the power of transmitter, modulation, bandwidth and etc. This information is printed in the ITU – Weekly Circular. So, the other operators, who think that such parameters will harm their services, have 4 months time to deliver their reclamation. In case there is one or a few operators, who are not satisfied about delivered characteristics of new comer, they can directly discuss or through ITU moderator discuss the problem and find the salvation suitable for both. Quite often such a decision is to reduce the sending power or the position relocation of the new operator. Only afterwards follows the so called Notification step, where ITU considers the characteristics of new comer and ratifies them if find correct. ITU acts as a guardian for the new operator against the other comers [cp.

Dodel99].

2.3   Types of satellites

Nowadays there are hundreds various types of satellites used for different services. They are categorized in the following types:

- Communication satellites

Their purpose of them is to serve as a relay station in the space using radio frequency waves to transmit the signal and information with it.

- Navigation satellites

The radio line signals sent from navigation satellites with the help of regularly developed electronic equipment enables the signals receiver on the earth to identify its position with pretty high accuracy.

- Earth observation satellites

These satellites are constructed with the goal to observe the earth from the space in order to monitor environmentally, make maps, use for meteorology, but usually not for military purposes.

- Astronomical satellites

The galaxies, other planets and other space bodies can be tracked and studied with the help of these satellites.

- Reconnaissance satellites

They are similarly as Earth Observation satellites are also used to watch the earth, but for military and intelligence (e. g. espionage) purposes. Governments do not provide much information about the power of these satellites as it used for various secret purposes.

- Solar power satellites

They use the radio frequency waves to transmit the power of sun to a huge antenna on the earth. The solar power afterwards can be used as a resource instead of traditional power.

- Space stations

The purpose of man shaped space stations is to create an environment for more and longer different scientific researches in comparison with other spacecrafts to measure the effects for human beings of a longer stay in the space.

- Weather satellites

Space vehicles are used to observe the weather and in some case the global climate.

- Miniaturized satellites

These satellites have uncommonly light weight and are very small (e.g. 500 – 10 kg compared with traditional satellites, which can weight about 5000 kg, like PAS 1 – R made by PanAmSat Corp.). The advantage of such spacecrafts is the much lower requirements for equipment in order to launch them into space, which leads to much lower costs. Besides that, they are also used for the missions, which usual satellites are not able to execute, like the low data rate transmission constellations, inspection of traditional space vehicles and etc.

- Biosatellites

In order to conduct the scientific tests and various experiments with the different living forms, the biosatellites were created.

- Killer Satellites

They are also named Anti-Satellite Weapons and are used for destruction of rival satellites or other weapons in orbits [Wiki06b].

2.4   Satellite orbits

In general, orbit is described as a pathway, which one space body makes around the other space body, because they are both influenced by gravity and centripetal force [Wiki06a]. The orbits, where satellites are launched by rockets, differ according their altitude above the surface of Earth and are most often categorized into the following classes (Figure 1):

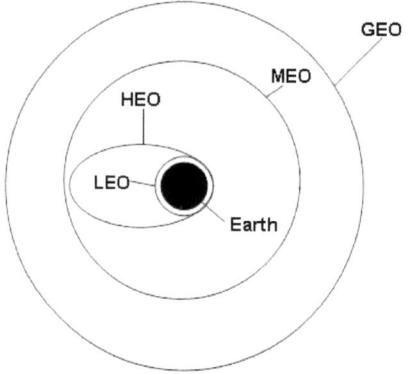

Figure 1: Satellites orbits [Walke00]

- Low Earth Orbit (LEO)

LEO finds its place from 200 km to 1200 km height above the earth. The advantage of this orbit is the shorter signal traveling time and lower possibility to loose its path. On the other hand, the coverage zone is quite small (in comparison with GEO) and the connection to satellite from ground station time is shorter, because the satellite moves quicker as the earth is turning. The increased interest in mobile communications via satellites over the last years motivated the augmentation of LEO usage and development of them.

- Medium Earth Orbit (MEO)

MEO is located between 1200 km and 35286 km altitude above earth surface. Some literature sources indicate that the Medium Earth Orbit is located between 5000 km and 13000 km height or between two Van Allen belts [Walke00]. Van Allen belts are two high intensity radiation zones of the earth, where highly charged particles and high energy neutrons take place. For this reason, the two belts are communication satellites damaging. Thus, it is avoided to place the satellite in the Van Allen belts zones.

- Highly Elliptical Orbit (HEO)

The name of the HEO arise form its elliptical form, which is helpful in order to achieve a better coverage of higher populated zones or usually not reachable parts of earth (such as poles) without the interruption of lower orbits [Walke00].

- Geostationary Orbit (GEO)

GEO is placed 35786 km above Earth's surface. The orbit is called geostationary orbit, because satellites', placed in this orbit, speed is matched with earth turning speed so that the satellite moves always together with the earth. In other words to say, if the one would be able to see the satellite from the earth, the satellite would always stay in the same point of the space from the earth perspective. Most of the communication satellites are place in GEO.

2.5 The advantages and disadvantages of satellite communication

The communication satellites play the middleman role between senders and receivers and operate as a relay station above the earth, which transmits the high radio frequency waves in order to deliver messages all over the world. The information passed from one place to another includes telephone conversations, pictures, TV and radio broadcasting, recently connections to internet and other [Satcom]. It is very convenient to use satellites communication in many situations of life. Thus, the main advantages of using satellite communication are listed bellow.

- Coverage of bigger geographical areas
- Availability of infrastructure straight after the start of satellite
- Application in remote areas without suitable infrastructure (planes, ships)
- One-way distribution to unlimited number of users
- The miniaturization of satellite terminal in progress
- Costs independence from the distance
- Global networks
- Worldwide mobility
- Functionality after natural and men - caused catastrophes
- Linkage with terrestrial networks

As satellites are placed high above the earth, they are able to provide service for bigger areas of the earth, depending on their altitude and beam width. Right after launching the satellite, they can be used for relaying a signal. There is no need for extra equipment in order to meet the expanded coverage zone. The constellations of satellites provide the worldwide coverage, so that the global communication is very quick and easy. All the customers in the same coverage zone, even in very remote areas can enjoy identical service level and costs, because they depend not on the distance, but on the costs of handling the satellite and its all belonging parts [Satcom]. The signal sent from satellite receives the limitless number of users if they have the appropriate tools to receive it. The development of satellite antennas is progressing very quickly nowadays, so that there is no need anymore to have huge sized terminals in order to receive the certain amount of data. The satellites transponder would be able to transmit the signal even after the natural or man caused disaster. Some countries, such as Japan and USA (New Jersey, California) assigned their catastrophes communications per law to satellites. While using spacecrafts it is possible to interlink them with the terrestrial systems. As an example, the big television channels use satellite communication to transfer their programs to the head ends of cables and further to provide their services for customer using wire lines.

Despite all these benefits, which provide satellite communication, it is useful not in every life situation. The reason is the relatively long signal traveling time, which causes the delay in telephone conversations or longer process while using internet. The costs of launching satellite to the space are pretty expensive. Furthermore, there are so many satellites today, that their coverage zones overlap and might cause the interference while transmitting the data.

To conclude, before starting to buy the equipment for receiving or sending messages via satellites, the usage of space vehicles' communications should be measured and the potential advantages defined, compared with other means of communication.

3   Satellite working principles

The understanding of main satellites working principles is important in order to apply the satellite communication for various purposes. Hence, this chapter illustrates the simplified model, how the satellite communication functions and its main characteristics.

The communication via satellite is a very good example of wireless networks. Besides space crafts based wireless systems there are also the terrestrial ones. The main difference between these two is the location of transmitter: either in sky (space) or on the earth.

3.1   Frequency bands

Starting from the beginning, all the signal in satellite communications are transmitted via electromagnetic waves, which are created by electric and magnetic fields and make 90° angle with each other as well as the direction of their propagation as illustrated in the figure (Figure 2) bellow [Combas].

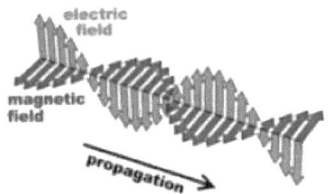

Figure 2: Electromagnetic wave [Combas]

Electromagnetic waves can be sent at different frequency bands, which is a certain range of frequencies, e.g. from 3.7 GHz till 6.5 GHz. The spectrum of frequency includes all possible frequencies from 0 till infinity. Radio frequency waves occupy the frequency range from 3 k Hz till 300 GHz and are used in satellite as well as some other communication forms. The information from spacecraft to the ground receivers and backwards is sent at the ITU classified particular range frequency bands in order to differentiate the signal as well as encoded information from uplink to downlink. The categories of radio frequency bands of waves are illustrated in the table (Table 1) bellow. Every range of radio frequency is assigned for a specific purpose, like military, marine use, navigational range, domestic use and etc. Some of the most commonly used bands are **S, C and Ku**, which are getting more and more overcrowded and for this reason the permission to transmit in these bands can be received not so easy. One more problem is that satellites are placed not so far from each other and the possibility of overlapping and interference is high, especially when spacecrafts use similar frequencies. In order to solve these problems, the limitations of signal strength and relay time were applied for these band ranges.

Frequency Band	Frequency Range (GHz)		Service
	Uplink	Downlink	
UHF	0,2 – 0,45	0,2 – 0,45	Military
L	1,635 – 1,66	1,535 – 1,58	Military / Navy
S	2,65 – 2,69	2,5 – 2,54	Broadcast
C	5,9 - 8,4	3,7 – 4,2	Domestic / Comsat
X	7,9 – 8,4	7,25 – 7,75	Military / Comsat
Ku	14,0 – 14,5	12,5 – 12,75	Domestic / Comsat .
Ka	27,5 – 31,0	17,7 – 19,7	Domestic / Comsat
SHF / EHF	43,5 – 45,5	19,7 – 20,7	Military / Comsat
V	- 60		Satellite Crosslinks

Table 1: Limitations of Frequency Bands Established by the ITU [Comsub94]

The higher the frequency, the higher is the possible data rate, but the more difficult the technology of satellite. All frequencies used in satellites communication are above 1 GHz, because for lower frequencies the atmosphere is not transparent so that the transmitting beam cannot get trough to the ground. The ITU organization as it was mentioned in the first chapters of this project is responsible for all the procedure of validation the selected frequency bands, which one satellite can operate.

3.2   Satellite communication parts

Two most significant parts of data relay via satellite are the **ground station** and **the satellite.** The transmitter of satellite is designed from the **uplink**, which is a ground-based element, and the **transponder,** which receives and reflects signals to the receivers and is placed on satellite [Satcom]. Multiple ground stations are distributed all over the world.

The main functions of satellite ground stations are receiving the information from the sender in form of electromagnetically waves (for example television program) and to pass it further to the satellite, which relay it to the target. For this reason, the ground stations have the full equipment, which is needed for the communication via satellite, such as antenna, transmitter, receiver, and coding as well as decoding device and other.

The modem, up converter as well as high – powered amplifier are the most important parts of usual uplink of the ground station. Customary, the modem transforms the signal of baseband coming from the users to the Intermediate Frequency (IF) (70 MHz or 140 MHz). Afterwards the upconverter is employed in order to translate the IF to high frequency RF waves. The last step is done at the High Power Amplifier (HPA), which amplifies the signal's power that it would be able to achieve the satellite.

Similarly the downlink is also combined of Low Noise Amplifier (LNA), Downconverter, and modem. Thus, the power of received signal in the ground station is increased using LNA. Afterwards follows the job of Downconverter, which change the RF waves to IF and over pass them to modem. In the latter step, adequately modem retransform the IF signals to the data for usage.

3.3   Data distribution

The table below illustrates the various data distribution scenarios using satellite communications. They differ from each other according the usage of satellites purposes (e.g. TT&C, data collection or data relay) and the number of stations to which data is spread (only to one point or broadcast to multiple number of participants). Tracking, telemetry and control means keeping the communication from the satellite and back to it as well as following and controlling the state of space vehicle from the earth [TTC06]. All the other descriptions are provided in the table.

Satellite Function	Data Dissemination Architecture	
○ Relay satellite ● Satellite Ground Station	Point – to - point	Broadcast
Tracking, Telemetry & Control	 TT & C data is transmitted between the satellite and ground station, either directly or by a relay satellite	 A relay satellite can provide broadcast TT&C service to multiple satellites.
Data Collection	 Satellite sensors collect data and transmit it to single ground station directly or by relay station.	 Satellite sensors collect data and broadcast it to multiple ground stations.
Data Relay	 Data - Relay satellites relay data originating from ground or from another satellite to single ground station.	 Data – Relay satellite broadcast data originating from ground or another satellite to multiple ground sectors.

Table 2: Data Dissemination Design Options [Comsub94]

3.4   The footprint of satellite

The satellite is capable to radiate the signal only to the areas of certain size. The zone, which satellite could translate is called the coverage zone or footprint. The size of the covered zone depends on the bright of its possibly greatest radiated beam and the height of satellite above the surface of the earth. It is calculated that in order to cover the entire world, the beam transmitted from antennas should be 17, 4° bright and the satellite should be placed in geostationary orbit. For example, as it is shown in the picture bellow, the three satellites operated by Intelsat provide the global coverage, which is illustrated as the bleached zones in every picture.

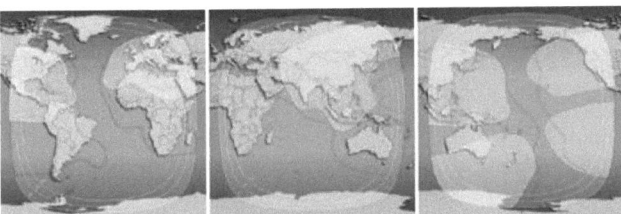

Figure 3: Global coverage by Intelsat [Satcom]

The very nice illustration of footprint zone of Eutelsat Hot Bird 5 satellites constellation where each of them are placed in different distance of the earth, but all at the 13° position of the East is provided in the picture below (Figure 4), which also clearly shows the coverage zone of satellites over the Europe.

Figure 4: The Eutelsat HotBird position at 13 Degrees East and the footprint [Satcom]

One of the most usually used antennas in the ground stations and satellites is the parabolic antenna, which consists of the horn and reflecting parabolic dish (Figure 5).

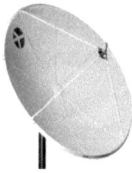

Figure 5: Parbolic antenna

With the help of such antennas the specific patterns of radiation or spot beams can be created in order to cover the desired regions and / or the high gain of antenna ensured.

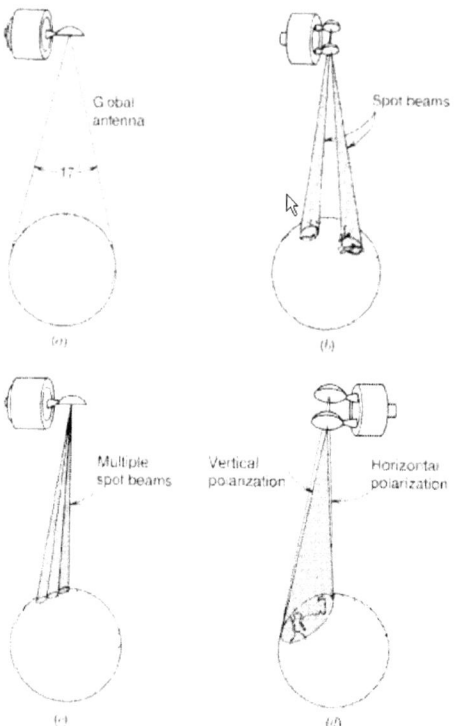

Figure 6: Satellite Antenna Patterns and Coverage Zones [Comsub94]

The picture (Figure 6) above illustrates the traditional footprints and behavior patterns of satellite antenna when the spacecraft is placed in geostationary orbit. The first one as it was already mentioned before with the 17,4 degrees width of beam provides a global coverage. The two spots beam covers certain regions in this case, because the operator would like to provide more services in these regions. The third scenario presents the multiple spot beams, which are radiated near each other and such positioning of beams provides two benefits: high gain antenna and relative big coverage zone. The last picture introduces two at the same frequency, but different polarization sent waves. The method used by satellite designer is called polarization when a part of satellite transponders sent electromagnetic waves to the earth in a vertical polarized form and the other part in horizontal polarized form. Nevertheless, the overlapping of two frequencies exist, they are not disturbing each other as there is a 90º difference between the phases. In such a way the capacity of satellites transmitter channels is increased as the transponder frequencies are reused, allowing to send much more amounts of information. One could say that using this, so called linear polarization, the same frequency can be used twice. When the polarization is like this the ground station (actually the receiving feedhorn of antenna) should be set up so, that would be able to differentiate vertically and horizontally sent signal at the same frequency. One more art of polarization is the circular, which is divided into right – hand and left – hand circular polarization used by other operators combines both electromagnetic waves together for sending a signal to the earth station. The different types of polarizations are shown in the picture (Figure 7) below.

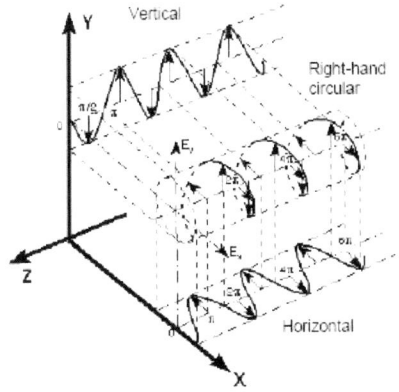

Figure 7: Different polarizations [Combas]

3.5   Signal Strength Decrement

The operators of satellites together with the coverage zones always provide the map (Figure 8) of radiated power, which helps for satellite communication users to set up the right size antennas in the right way in order to reach more power.

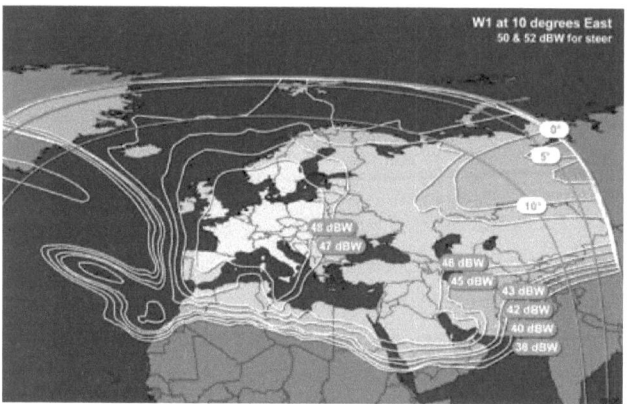

Figure 8: Eutelsat W1 regional coverage

The ground station transmits the signal to the satellite at the certain power ($P_{radiated}$). As the power capacity of the transmitted station is limited, the radiated power of the signal with the help of antenna's shape is increased. The parabolic antenna, which looks like a dish and has a feedhorn for focusing the transmitted and received signals, directs the radiation waves all to the same satellite direction, which compared to isotropic radiating antenna, delivers more powerful stream. The transmitted power density of isotropic antenna is equated to 1 in order to illustrate how much bigger the power density of other antenna is compared to isotropic radiating one. As an example, the dipole antenna has a G of 1,76 dB, which means, that the power density of dipole antenna is 1,76 times higher and consequently it means, that dipole antenna transmits and receives the signal beams with more power. The satellites and ground station antennas are also constructed so, that the gain would be high as possible. The $P_{received}$ at the satellite or ground station could be calculated according the formula bellow[1]:

---

[1] Source: lecture handnotes

$$P_{rec} = P_{rad} * G_{sent} * G_{rec} * [\frac{\lambda}{4\pi r}]^2$$

This equation demonstrates the direct dependence on the gains of receiving ($G_{receiving\ antenna}$) and sending ($G_{sending\ antenna}$) antennas as well as radiated power (P_{rad}) and conversely dependence on transmitting spacecraft distance above the earth (r). Actually, fully explaining this equation should be mentioned that the received power $P_{received}$ depends on the gain of sending antenna ($G_{sending\ antenna}$), the spherical area ($4\pi r^2$) and the effective antenna area ($\lambda^2/4\pi * G_{receiving\ antenna}$). As a matter of fact, the Gain factor of the same antenna gets weaker as the antenna radiation is reflected further from feed horn (from centre to the outer sides of antenna), which caused the decrease of received power getting from the middle to the sides of coverage zone on the earth. Furthermore the distance from the satellite to the earth gets longer as moving from the central part to the side, which impacts the weaker received power as well.

To summarize, the main factors to determining the different signal strength distribution over the coverage zone of satellite are the gains of both antennas (transmitting and sending) and the length of radius from satellite to the receiving station.

The amount of power received by antenna differs depending on antennas effective areas. In order to receive higher power signal the larger diameter and/ or higher gain antennas should be used as one factor of received power amount is the affective antenna area.

3.6   Link Budget

The signal transmitted from satellite to another ground station (downlink) is strengthened time to time using particular equipment, because it experiences various losses. The sum of all gains and losses used either to calculate the power needed to be transmitted from the satellite or the quality (power) of the signal received at the ground station is called link budget, which shows the performance of system at the very end of this process. The simplified model of link budget is going to be presented in this chapter. First of all the main components influencing the link budget will be described:

Pathway length (L)

The length of the way, which signal has to move from transmitter (spacecraft) to the receiver is defined as path length. This distance depends on how far from the satellite receiver is located and the size of the angle, which is created between the line drawn to the receiver and the shortest altitude to the earth line.

Beam width

The signal sent from satellite antenna creates a cone shape illumination directed to the earth. Thus, the beam width is the cone's vertex angle. For example, the vertex angle of the satellite located in GEO with the target to reach global coverage is 17.4º (as it was mentioned before). So, its beam width is 17.4º.

Aperture diameter

The decision about antenna's size is made according diameter of its aperture. Nevertheless the beam width is directly related to the aperture diameter. There could be different sorts of antennas used in order achieve the appropriate beam width and aperture diameter proportion.

Gain (G_t)

The gain of antenna was already described in the 3.6 chapter. Just to sum up, the antenna gain depends on its shape, the aperture diameter area, the efficiency of antenna and the length of signal wave. In case antenna is parabolic (as most antennas in satellites), it concentrates quite a big amount of reflected energy into the beam and sends it to the earth. The bigger size antenna can achieve the higher gain and narrower beam. As mentioned before, the beam strength and the gain of antenna is the most strongest in the middle of the signal and gets weaker moving to the sides (Figure 8).

Signal – to – Noise ratio (SNR)

The sent information has to reach the ground station without any damages or mistakes. For this reason, the receiver has to be setup so, that it could distinguish between the real signal and the noise appeared during its tour. The signal and noise ration demonstrates how much the signal is stronger as the noise and shows the quality of the signal at the receiving station. The determination of the SNR can be made with the help of the most wanted Bit Error Rate (BER), which refers to the possibility of getting the signal bits with errors. As an example, the BER of 10^{-6} (1 inaccuracy in one million bits) is quite usual for very precise data or high quality signal. When the BER is chosen to be 1 error per 100 bits (10^{-2}), than the SNR is low (a bit more as 2) and the extra 1 to 3 dB power should be added in order to correct the errors.

Line loss (L_l)

The power lost between the transmitter and antenna is named the line loss. According the made estimations, the line loss is usually -1 to -3 dB.

Space loss (L$_s$)

This parameter describes the power loss during the signal travels from sender to the receiver and depends on the signal's way length, described in the section above as well as frequency of the signal and light speed.

Antenna loss or attenuation loss (L$_a$)

The loss of antenna also reduces the power of transmitted signal. The main reason for antenna loss is so called rain attenuation, which is depending on the wave's frequency, satellites elevation angle, rain drops size and etc. Thus, the loss of antenna is defined mainly according rain attenuation.

Noise temperature (Ts)

The signal is getting weaker overall during its travel, because of many different noises appearing in sending antenna, the downlink line and receiver. The noise temperature converts all these effects into one quantity, which illustrates the loss of energy as a heat.

Receiver gain (G$_r$)

It is defined the same as the transmitted antenna gain. Just the efficiency of antenna usually is higher as engineers steady improve the quality of antennas in ground stations [Comsub94].

Link budget equation

To sum up, the simplified equation to calculate the quality of the received signal is provided below:

$$SNR = P + G_t - L_s - L_a - L_l + G_r - 228.6 - Ts \ [$$

Dodel99]

P = transmitted power (dBW)

SNR = Signal – to – Noise ratio (dB)

L$_l$ = line loss (dB)

G$_t$ = transmitting antenna gain (dB)

L$_s$ = space loss (dB)

L$_s$ = attenuation loss (dB)

G$_r$ = receiving antenna gain (dB)

Ts = system noise temperature (K)

228.6 dBW/ K = calculated coefficient of loss in all transmission cases

The equation to calculate the amount of transmitted power in order to receive a certain quality of the signal can be derived from this formula.

Calculation of link budgets allows the better decisions while choosing the right transmitter with the right mass and right input power.

4 Application

The possible spheres where the one could experience the advantages of satellite communications are described in this chapter.

4.1 Television and Radio

As just mentioned, satellite communication provides considerable number of advantages. Nevertheless, the usage of them is not worth in every communications activity. Hence satellites are in the first place used to broadcast television and radio. These two services differently from telephony have the immune for signal delay, because the signal is transferred at the same sequence as it was sent and the quality of view or sound does not suffer. The two kinds of satellites are used to broadcast television and radio signals: Fixed Service Satellite (FSS) and Direct Broadcast Satellite (DBS). The two definitions of satellites differ a bit in North America and Europe depending on the transmitting technology they use. The Direct – to - Home (DTH) services are provided by broadcasters using DBS technology. The simple usage and relative cheap usage of DTH service plays the significant role in growing number of its customers, who can buy and install the equipment easily by themselves or with the help of technicians in order to receive the programs. A number of different possibilities can be offered to the customers by programs suppliers. All the viewers can receive free-to-air (FTA) channels. However, the restrictions are added on some type of programs designed to a certain public or in other word programs are encrypted. On the purpose to watch the channel, the encryption should be unscrambled with a special device, which typically is integrated in the receiver and accordingly named Integrated Receiver Decoder (IRD). The satellite TV provides the large scale of available programs from the entire world in comparison with the cable TV, which is usually local oriented except some channels which are broadcasted by satellites to the cable TV head ends and offered for viewers. As today's world is getting more and more digitalized, the signal transmitted via satellites is also changing from analogue to digitalized, where the sent data is compressed allowing to relay the huge number of channels over one transponder. In comparison the analogue by itself is using one full transponder, which needs a significant bandwidth of 36 or 72 MHz. The most digital signals used today are converted to MPEG-2/DVB format, which can be received by the digital reception tool, unscrambled and than the full data flow can be divided into different types of information. Generally speaking, the developed of digital technology provides many benefits to the consumer, because he can better control the content of delivered information, obtain increased quality data (audio, video) and the selection of programs and service is enhanced.

The satellite communication is used broadly today for business video conferences, which allows making common meetings with the adequate persons all over the world at the same time. The centers of remote learning are using satcom as the lecture can be translated to different students everywhere they are [Satcom].

4.2   Telephone Service

One more field, where the satellite communications are used is telephone service, but up to date only in areas where there is no suitable infrastructure for cable or other technologies, such as planes, ships, desserts, distant islands and etc.

Figure 9: Satellite and cellular telephones

The mobile or cellular phones are not based on satellite communications as well. Instead they are using ground based translating and receiving fleets. Similarly, most part of the international telephone conversations today are also executed not with the help of satellites. First of all, using optic fiber is very convenient as the data rate, which can be send at once is very high and speedy. The connection without delays can be established to almost every continental place as in contrast the signal sent via satellites based would be always with a few seconds delay. The reason for that the long distances the signal has to travel (from transmitting station to satellite and back to receiving station) as well as the dependence of transmitted beam of the speed of light ($c = f * \lambda$), which is also limited.

4.3   Internet Service

Similarly as in telephone services, recently the technology of satellite communications was started to use for the Internet access "via broadband data connection" in remote areas, where the conventional (wire line or dial up) connections are impossible. For example, the end-user in remote island who wants to get the some information, first send a request for information, which goes trough cable via ground station or trough his antenna straight to satellite, than to the receiving ground station and from it to the broadband internet provider. The requested data are sent all the ways back till receives the end-user).

The both above described satellite services cost more than the traditional ones, because of the higher technology, such as special developed satellite telephone antenna, is required (Figure 9).

4.4   Very Small Aperture Terminals

Furthermore, possibly the most rapidly growing area of satellite communication technology since 1980's is the use of very small aperture terminal ground stations (VSAT), which rapidly started to be installed in shops, banks as well as other business units and used for so called point – of sale. The various financial transactions, inventory control, the entry of order and further related applications, could be performed using VSAT. The already mentioned delay of the signal is not as important in such operations as contrariwise in telephone conversation. For example, the 1 second delay during the financial transaction in a drive – in bank is almost not possible to notice. Therefore, the application of satellite communication is growing significantly for all the services, where the delay of information is not critical [Held91].

4.5   Other applications

The satellite communication is also used for Fax services. Furthermore, Telex services are also based on satcom, which are mostly used for maritime such as Navtex assigned to deliver to ships forecasts or warnings about navigational and meteorological changes as well as information relative safety of the ship. The High Volume Trunk, like international data transportation could be also made via satellites. The usage of satellite communications was adjusted in industries such as press, where centrally created contents of newspapers or magazines are sent via satellites to local print factories, what enables the citizens to receive the brand new copy of the edition early in the morning including the news from last day.

5 Conclusion

Satellites and their systems play the significant role in life of all world population as the enormous possibilities were created together with this space orbiting machine. After the first space vehicle Sputnik 1 was created based on the innovative ideas of Mr. A. Clarke, the quick development of satellites started.

There are many different spheres, where the satellites and their fleets can be used today. The maps and weather forecasts can be created, the position of different objects defined, the other planets and galaxies can be observed, the scientific researches made, the secret military and usual private information can be sent with the help of satellites.

Despite of a few disadvantages (e.g. signal delay) and higher costs of launching satellites into the space, the overall advantages of using satellite communication in certain spheres is irreplaceable today (e.g. catastrophes places). Especially the communication satellites serve the remote areas, such as planes, ships and distant islands.

With the help of satellites, the entire surface of the world can be covered, which provides the speedy communication all over the world. Besides the transmitted signal decrement moving from center to the outer parts of coverage zone, the power of the signal can be increased using appropriate gain and size antennas, which means the service provided is equal to all users. The various sizes signal beams directed to different places can provide the global, regional and spot coverage, depending on which region the satellite operator is focused on.

The satellites, ground stations and sending as well as receiving antennas are regularly developed and improved, what lets to assume that future usage of communication satellites as well as other types will increase even more providing us new services and products.

List of Literature

Dodel99	Dodel H.: Satellitenkommunikation. Heidelberg: Hüthig Verlag, 1999, P. 25 – 28
Antenna06	http://www.irte.it/consumersat/csap240cmuk.htm, accessed 25.06.2006
Held91	Held G.: Understanding data communications. Wiltshire: Redwood Books, 1991, P. 536 – 539
Walke00	Walke B.: Mobilfunknetze und ihre Protokolle. Band 2. Stuttgart, Leipzig, Wiesbaden, B. G. Teubner GmbH, 2000, P. 433 – 493
Wiki06a	http://en.wikipedia.org/wiki/Orbit, accessed 23.06.2006
TTC06	http://ocw.mit.edu/NR/rdonlyres/Aeronautics-and-Astronautics/16-851Fall2003/6AC09E33-FD3D-4C85-899F-DDBF14109E8A/0/l20_satellitettc.pdf, accessed 22.06.2006
Comsub94	Fischer M.: Communication Subsystems, 1994 http://www.tsgc.utexas.edu/archive/subsystems/comm.pdf
Combas	Satellite Communications Basics http://www.panamsat.com/customer_support/satellite-communication-basics.pdf, accessed 12.06.06
Wiki06b	http://en.wikipedia.org/wiki/Satellite, accessed 14.06.06
Wiki06c	http://en.wikipedia.org/wiki/Satellite_communication, accessed 14.06.06
Glossary	http://www.satnews.com/free/glossary.html#V, accessed 16.06.06
Satcom	http://www.jisc.ac.uk/sat_report3.html, accessed 18.06.06